Mind Hacking

Highly Effective Ways to Smash Unhealthy Mindsets

(a Simple Plan to Maximize Your Potential & Tap into Your Inner Confidence)

Joshua Miller

Published By **Hailey Leigh**

Joshua Miller

All Rights Reserved

Mind Hacking: Highly Effective Ways to Smash Unhealthy Mindsets (a Simple Plan to Maximize Your Potential & Tap into Your Inner Confidence)

ISBN 978-1-7382957-0-8

No part of this guidebook shall be reproduced in any form without permission in writing from the publisher except in the case of brief quotations embodied in critical articles or reviews.

Legal & Disclaimer

The information contained in this book is not designed to replace or take the place of any form of medicine or professional medical advice. The information in this book has been provided for educational & entertainment purposes only.

The information contained in this book has been compiled from sources deemed reliable, and it is accurate to the best of the Author's knowledge; however, the Author cannot guarantee its accuracy and validity and cannot be held liable for any errors or omissions. Changes are periodically made to this book. You must consult your doctor or get professional medical advice before using any of the suggested remedies, techniques, or information in this book.

Upon using the information contained in this book, you agree to hold harmless the Author from and against any damages, costs, and expenses, including any legal fees potentially resulting from the application of any of the information provided by this guide. This disclaimer applies to any damages or injury caused by the use and application, whether directly or indirectly, of any advice or information presented, whether for breach of contract, tort, negligence, personal injury, criminal intent, or under any other cause of action.

You agree to accept all risks of using the information presented inside this book. You need to consult a professional medical practitioner in order to ensure you are both able and healthy enough to participate in this program.

Table Of Contents

Chapter 1: Understanding The Power Of Manipulation 1

Chapter 2: Mastering Emotional Intelligence 13

Chapter 3: Hacking The Habit Loop 25

Chapter 4: The Power Of Positive Affirmations 40

Chapter 5: Harnessing Flow States 56

Chapter 6: Embracing Constructive Discomfort 72

Chapter 7: The Gratitude Advantage 83

Chapter 8: Organizing Thoughts For Clarity 97

Chapter 9: Social Connection The Neurological Boost 104

Chapter 10: From Manipulation To Mastery 111

Chapter 1: Understanding The Power Of Manipulation

Manipulation, at its middle, is a tool. Like any device, it is able to be used for actual or sick, to build or to interrupt. To a few, the very word brings up pictures of deceit, control, or possibly harm. But proper proper right here lies an omitted fact: manipulation, even as directed inward and used ethically, has the capability to result in profound personal growth.

Let's take a moment to recognize what manipulation in reality approach. In a large revel in, manipulation is ready converting something to gain a favored final effects. Think of a sculptor working with a piece of clay. They take a look at stress, shape it, and mildew it until it turns into a beautiful statue. That's manipulation. But in this situation, it is a splendid and progressive stress.

When it involves the thoughts, the identical standards practice. The human thoughts are malleable, similar to that piece of clay. It can be customary, added approximately, and informed to reap diverse results. Ethical manipulation of the mind is not approximately deceit or manage over others, but approximately mastery over oneself. It's about shaping mind, feelings, and conduct to foster growth and bring about a brighter future.

Consider a smooth instance. Jane continually believed she end up awful at math. This notion came from a unmarried terrible grade she acquired in fifth grade. Since then, she shied some distance from numbers, glad of her private disability. However, subsequently, Jane decided to undertaking this notion. Instead of telling herself she end up lousy at math, she began telling herself she certainly preferred extra practice. She began to cope with math problems every day. Slowly, she started out improving. By manipulating her very non-public belief about herself, she grew to emerge as a perceived vulnerable element right proper into a power.

Just as Jane reshaped her beliefs, so can others harness the strength of manipulation for self-improvement. But it's far crucial to apprehend the limits of moral manipulation.

Ethical manipulation specializes in the self. It's about understanding private weaknesses, strengths, beliefs, and behaviors, and then tweaking them for the higher. It's not approximately controlling or influencing others. It's approximately self-mastery.

For moral manipulation to be powerful, it must be based totally mostly on honesty. This way acknowledging personal flaws without judgement and spotting areas for development. It's not about growing a fake experience of self or pretending to be a person else. Instead, it's miles approximately optimizing the prevailing self.

For example, recall a person who has a bent to procrastinate. They frequently cast off duties until the very last minute and then rush to finish them. This behavior consequences in strain and subpar paintings. Through ethical manipulation, this person have to apprehend the foundation motive in their procrastination – probably it is fear of failure or a dependancy advanced at some point of university days. By information the cause, they'll be able to artwork on converting the conduct. Maybe they begin by way of breaking obligations into smaller steps or placing earlier final dates for themselves. Over time, this manipulation can result in better paintings behavior and plenty a lot much less stress.

But, it's essential to recall that alternate does no longer happen overnight. Just much like the sculptor who should mold the clay time and again to get the popular form, so want to people paintings continuously to look results. It's a machine of trial and errors, of information what works and what could no longer.

In the grand scheme of things, ethical manipulation is a path to empowerment. It's a focal point that one isn't best a made from situations but can be an architect of their very very very own future. By taking price of personal beliefs, emotions, and behaviors, you probably can direct their adventure closer to increase and greatness. Just because the sculptor sees capability in a lump of clay, so ought to people see functionality in themselves.

And with the right manipulation, that capacity can be discovered out.

In prevent, manipulation is a powerful tool. But like every tool, its impact is based upon on how it's far used. Directed inward and used ethically, it could be a stress for superb fantastic alternate. It can assist rework beliefs, shape behaviors, and guide humans in the course of a destiny filled with boom and achievement. By records and harnessing this energy, it is straightforward to pave the manner for a journey of self-improvement and greatness.

The Self-Dialogue Shift

Every man or woman consists of with them a normal partner: their inner voice. This voice is an ongoing commentary on existence, a bypass of mind, ideals, and feelings that shape how someone sees themselves and the area around them.

This internal narrative plays a powerful position in influencing actions, reactions, and selections. For higher or worse, it is the tale instructed to oneself, approximately oneself.

The problem arises while this narrative is bad. Imagine someone looking for to learn how to adventure a bicycle. If their inner voice continues pronouncing, "It's too difficult. I'll in no manner get this right," the chances of them giving up are high. On the opportunity hand, if the voice says, "It's hard, however I'll get there with exercise," they may be much more likely to persevere and ultimately succeed.

Changing this internal narrative, moving the self-communicate from bad to excessive top notch, will have profound effects on non-public boom and capability.

Think approximately Sam. From a younger age, Sam become knowledgeable he turn out to be clumsy. He believed this narrative, and it have become part of his self-speak. Every time he'd drop some issue or journey, he'd assume, "There I skip once more, being my ordinary clumsy self." This belief held him all over again in masses of regions of existence, especially sports activities activities sports. He turned into too scared to try for worry of confirming his 'clumsiness.'

However, a shift took place while Sam joined a dance class on a whim. Initially, his self-communicate was the equal. He'd skip over a step and right now don't forget his 'herbal clumsiness'. But as schooling went on, his instructor started out out to offer him high-quality comments, noting that he had precise rhythm and coordination. Slowly, Sam began to question his vintage notion. He commenced out to update "I'm so clumsy" with "I without a doubt need greater exercise" or "Everyone makes errors sometimes."

Over time, not best did Sam emerge as a in a position dancer, but his new self-communicate additionally started out out to influence one in all a kind regions of his lifestyles. He attempted out for sports sports he'd in no way taken into consideration in advance than and located he was pretty true at a number of them. The shift in his narrative unfold out a global of opportunities.

So, how does one start this way of shifting the self-talk?

The first step is recognition. It's crucial to apprehend and be aware of that inner voice. What stories are being knowledgeable? Are they useful or risky? Do they push closer to growth or keep decrease again?

Once this attention is mounted, mission those terrible narratives. Ask questions like, "Is this virtually real?" or "Is there proof to assist this belief?" Often, upon mirrored image, many will find that their self-talk is primarily based mostly on antique ideals or isolated incidents that do not define their whole being.

Next, replace the terrible with the remarkable. For every horrific declaration recognized, offer you with a counter brilliant announcement. For instance, if the concept is "I'm no longer clever sufficient," counter it with, "I clearly have the functionality to investigate and broaden."

Practice is critical. Just as a muscle desires repeated exercise to increase more potent, the mind desires everyday exercising to cement a state-of-the-art narrative. Every time a terrible idea creeps in, apprehend it, challenge it, and update it.

To help with this technique, it might be useful to maintain a magazine. Write down terrible mind and beliefs and then counter them with nice ones. Over time, this workout can help decorate the current narrative.

Consider the story of Lisa, a young author. She had a sturdy choice to install writing a very unique, but her self-speak end up whole of doubt. Thoughts like "I'm now not correct sufficient" or "No one will want to take a look at what I write" plagued her. She decided to task this narrative. Every time she had a horrible belief, she'd write it down after which counter it with a nice one. Over time, her magazine have end up a testament to her changing self-communicate. It was full of statements like, "Every writer starts offevolved offevolved somewhere" and "I write because of the truth I adore it, no longer for others." Eventually, she wrote that novel, and it modified into a mirrored picture of her newfound self perception and perception in her.

The electricity of the self-talk cannot be overstated. It's the story of 1's life, narrated via oneself. By becoming aware of this narrative, hard it, and changing negativity with positivity, absolutely everyone can shift their trajectory. This shift can open doorways, harm obstacles, and pave the manner for boom and greatness. The pen is in hand; it is time to rewrite the tale.

Chapter 2: Mastering Emotional Intelligence

Emotions color critiques, stress moves, and, often, determine fulfillment or failure in numerous ventures. Therefore, statistics and coping with those feelings becomes paramount for everybody looking for personal increase. This statistics and control of feelings is often termed as "emotional intelligence."

At its center, emotional intelligence is the capability to recognize, understand, and manipulate not exceptional private feelings however moreover the feelings of others. It's approximately being tuned into the emotional channel of the arena, selecting up on diffused cues, and navigating situations with each empathy and willpower.

Take Alex, as an example. Alex labored in a crew wherein tensions regularly ran excessive because of tight remaining dates. While others might also need to often get flustered or upset, Alex had a knack for staying calm. He can also need to select out up on his teammates' frustrations and generally knew the proper factor to say to ease the state of affairs. He wasn't really fending off war; he come to be analyzing feelings, knowledge them, and responding in a manner that benefited clearly all people. This is emotional intelligence in motion.

So, why is reading emotional intelligence important for non-public boom?

Firstly, feelings play a feature in almost every choice made. When a person is privy to their feelings and might manage them, they're higher ready to make sound choices. Acting out of anger, as an example, frequently leads to regret. But recognizing that anger, know-how its supply, and selecting the way to answer can alter the direction of sports activities in a greater top notch course.

Secondly, relationships, both private and expert, thrive on emotional intelligence. Understanding and respecting the feelings of others ends in stronger, more big connections. It's the bridge that links human beings together, bearing in mind deeper verbal exchange and cooperation.

How, then, does one move approximately studying emotional intelligence?

Self-consciousness is the location to start. Before understanding others, one ought to first understand oneself. This technique being in music with personal feelings, spotting them as they rise up, and statistics their motives. It's about asking questions like, "Why am I feeling this way?" or "What brought on this emotion?"

For example, consider Jamie, who noticed she'd get irritated each time her colleague, Mark, spoke up in meetings. Instead of brushing the feeling aside, she dived into it. She discovered out it wasn't Mark's mind that bothered her but his tone, which reminded her of a faculty bully from her beyond. By recognizing and information this emotion, Jamie have grow to be higher prepared to manipulate her reactions in destiny conferences.

After self-interest comes self-regulation. It's one thing to apprehend an emotion and a few other to determine a way to behave on it. Self-regulation is ready taking a second in advance than reacting. It's considering the first-class path of action in place of being swept away via the initial emotional response.

There's moreover a need to domesticate empathy. Empathy is the functionality to region oneself in some unique's footwear, to enjoy what they feel. It's a step past certainly recognizing a person else's feelings; it's certainly information them. When someone can empathize with others, they might better navigate social situations, offer help, and assemble more potent relationships.

For example, consider Ryan who determined his friend Ben looking downcast. Instead of brushing off it or developing a mild-hearted commentary, Ryan approached Ben, requested him how he felt, and virtually listened. By empathizing with Ben's situation, Ryan turn out to be no longer most effective capable of provide comfort but furthermore assist their bond.

Lastly, studying emotional intelligence requires social capabilities. These are the gear that allow a person to interact harmoniously with others. It's approximately effective conversation, battle choice, and building bonds.

It's well nicely worth noting that mastering emotional intelligence isn't always approximately suppressing feelings. Emotions are herbal and valuable. They provide insights into the place and the self. Instead, it's far about harnessing them, guiding them in procedures that cause outstanding outcomes.

Emotions are like a river. They can be calm and serene, or wild and tumultuous. Without information and manipulate, this river can overflow, causing chaos. But with emotional intelligence, it may be channeled, directed, and harnessed for non-public boom. The adventure to studying emotional intelligence is a journey to a extra balanced, information, and successful self.

The Art of Visualization

Every invention, every achievement, each bounces in advance started as a notion. These thoughts, while paired with purpose and creativeness, can emerge as incredible photographs that manual actions and alternatives. This manner, called visualization, is an artwork that has been practiced for centuries with the resource of the ones aiming for greatness.

Visualization is more than mere having a pipe dream. It's an intentional act of crafting and molding the destiny thru using growing a smooth and specific intellectual photo of a favored final effects or reason. It's approximately experiencing the future on the identical time as however inside the gift.

Consider the story of Alicia, a younger athlete with dreams of triumphing a gold medal in the Olympics. Every day, as part of her education routine, she may additionally need to close to her eyes and remember herself on the Olympic stadium. She can also want to pay interest the roar of the group, revel in the weight of the medal round her neck, and note the proud smiles of her own family within the stands. This extremely good picture stored her advocated, driven her limits, and, through the years, played a essential characteristic in her attaining that very dream.

So, what makes visualization this form of mighty device?

Firstly, the mind has a completely specific feature. It regularly struggles to distinguish amongst a vividly imagined event and a actual one. When an event is visualized in element, the thoughts reacts as though it is taking region in truth. This technique that the emotions, the emotions, or maybe some bodily reactions expert sooner or later of visualization can be nearly as immoderate as experiencing the real occasion.

This reaction has profound effects:

1. Preparation: Visualization prepares the thoughts and frame for the real occasion. If a person continuously visualizes themselves giving a a achievement presentation, for instance, at the same time as the time includes deliver it, they've got already "been there" generally in their thoughts. This familiarity reduces tension and boosts self belief.

2. Motivation: Creating a easy and compelling photograph of the future offers a powerful incentive. It turns into a beacon, guiding actions and selections within the direction of that favored future.

For powerful visualization, clarity is important. It's no longer sufficient to have a indistinct photo. The more distinct and particular the visualization, the stronger its outcomes.

Think of Michael, an aspiring musician. Instead of in truth imagining himself playing tune, he visualized every element. He may additionally need to peer the kind of venue, the people within the crowd, the lighting, or even the garments he was wearing. He may want to pay interest the appropriate songs he changed into playing and experience the feelings they evoked. This top notch visualization no longer super inspired him but additionally gave him direction. He knew exactly what he desired and could tailor his efforts toward that imaginative and prescient.

However, visualization isn't pretty an awful lot seeing. It engages all of the senses. It's approximately feeling the feelings, paying attention to the sounds, or maybe smelling the scents associated with the imagined scenario. Engaging a couple of senses makes the visualization more great and impactful.

Moreover, consistency is top. Just like a few exclusive talent or art work, the electricity of visualization grows with exercise. The more regularly and frequently it's miles practiced, the clearer and greater influential the visualizations grow to be.

It's additionally nicely well worth noting that at the identical time as visualization is a powerful device, it is not a possibility for motion. It's a complement to it. Visualization gadgets the diploma, presents the incentive, and prepares the thoughts and body, but to carry the imaginative and prescient to fact, tangible steps and efforts are essential.

The destiny, on the identical time as uncertain, is moldable. With the artwork of visualization, you probably can craft and form the trajectory of existence. Through intentional imagination, the mind may be educated, added approximately, and organized to show dreams into fact. Visualization is not pretty an awful lot seeing the destiny; it's miles approximately developing it.

Chapter 3: Hacking The Habit Loop

In existence, behavior play a prime characteristic. These automatic actions and reactions shape days, impact choices, and over time, mould destinies. While some behavior propel humans closer to success, others can preserve them decrease decrease lower back. Hence, know-how the internal workings of behavior becomes crucial for virtually really every body aiming to harness their complete capability.

Habits aren't random. They follow a predictable cycle, regularly referred to as the dependancy loop. This loop consists of three stages: the cue, the habitual, and the praise.

To start, there may be the cue. It's the trigger that initiates the addiction loop. Cues can be outdoor, like a ringing mobile phone signaling an answer, or inner, like a sense of boredom leading to snacking.

These cues are indicators to the mind to begin a specific every day.

Following the cue comes the ordinary. This is the motion or conduct related to the addiction. When the mobile telephone earrings, the ordinary is probably choosing it up. When feeling bored, the everyday might be starting off the refrigerator. It's the core motion of the dependancy loop.

Finally, there's the praise. It's the terrific outcome or feeling due to the routine. Answering the cellphone can also moreover result in a pleasing verbal exchange, a reward for the movement. Eating a snack may additionally alleviate boredom, presenting a sense of delight.

Consider the case of Sam, who, each afternoon, located himself munching on cookies. The cue have become the afternoon stoop, that period after lunch wherein energy degrees dip. The habitual become consuming cookies, and the reward have turn out to be the sugar rush and the short improve in mood. This loop saved repeating itself each day, critical to weight advantage and health issues for Sam.

So, how can the information of the addiction loop be leveraged for powerful alternate?

The first step is recognition. Before a dependancy may be modified, it must be diagnosed. It's approximately identifying the cues, routines, and rewards in cutting-edge behavior. Sam, as an instance, had to apprehend that his afternoon stoop turned into maximum vital him to are looking for for out sugary treats.

Once a dependancy is identified, the following step is dissection. This includes breaking down the dependancy loop and studying every element. What is the cue?

Why does this specific recurring comply with the cue? What praise is being sought? By expertise the ones factors, techniques can be unique to regulate or update the addiction.

With this know-how, the zero.33 step is substitution. The only way to alternate a dependancy isn't to dispose of it but to replace it. The cue frequently stays the same, because it's hard to trade triggers, particularly if they're outside. The normal, however, may be swapped. Instead of accomplishing for cookies, Sam need to pick out out to take a brief walk or have a cup of inexperienced tea. The key is to find a new ordinary that gives a similar praise. In this example, the reward is a harm from the afternoon monotony and a lift in electricity.

Consistency is the very last piece of the puzzle. Changing conduct is difficult, and there might be times of regression. But with persisted try and know-how of the dependancy loop, antique patterns may be changed with new, greater useful ones.

One important element to recall is that now not all behavior need to be modified. Many conduct are useful. The aim isn't always to put off conduct but to make sure that they align with desired effects and goals.

In the grand scheme of private boom, conduct act because of the fact the constructing blocks. Each addiction, no matter how small, contributes to the larger shape of lifestyles. By hacking the dependancy loop, facts its components, and leveraging this know-how, you can gather a existence of reason, reason, and greatness. Whether seeking to domesticate new conduct or exchange present ones, the cue-regular-reward cycle offers a roadmap to success.

Mindful Meditation: The Gateway to Presence

Mindful meditation, at its middle, is prepared statement without judgment. It's the paintings of anchoring the mind to the modern 2d, experiencing it clearly with out letting it get colored via the usage of manner of biases or beyond tales.

Consider the tale of Nina, a pupil crushed through the stresses of university and life. Her days have been fed on with issues approximately upcoming tests, regrets over beyond errors, and anxieties approximately the future. One day, a teacher introduced her to the workout of aware meditation. Nina positioned to sit quietly, focusing on her breath and letting her thoughts glide without getting associated with them. Over time, she located that she have to concentrate better, her problems appeared much less daunting, and she or he or he felt greater associated with the sector spherical her.

So, why does conscious meditation have such a profound effect?

Firstly, the workout brings about a heightened feel of self-recognition. By focusing on the breath or one-of-a-kind focal elements, one will become acutely privy to the inner panorama of the mind. Thoughts are observed, however in choice to reacting to them, they'll be clearly positioned. This non-reactive statement permits for a deeper information of one's concept patterns, feelings, and reactions.

Secondly, conscious meditation enhances recognition. In a worldwide riddled with distractions, keeping hobby may be hard. However, the exercising of returning one's interest to the breath or a designated anchor, over and over, hones the skills of recognition. Just like a muscle strengthens with workout, the mind's capacity to pay interest improves with consistent meditation.

Additionally, this shape of meditation cultivates a revel in of equanimity. Life is filled with u.S.A. Of americaand downs, joys, and sorrows. By training non-judgmental statement, one learns to approach all reviews with a balanced mind. Rather than getting overly excited with the aid of using positives or overly dejected via negatives, a regular, even-tempered response is nurtured.

To illustrate this, don't forget a pond. When stones (thoughts or memories) are thrown into it, ripples are created. But with time, the pond returns to its calm nation. Similarly, the thoughts, with everyday aware meditation, can discover ways to skip lower back to a kingdom of calm and balance, no matter the disturbances.

To embark on the journey of conscious meditation, no tough arrangements are favored. It can be as easy as locating a quiet spot, sitting with out problem, ultimate the eyes, and turning the attention inward. The breath serves as a reachable anchor. As one inhales and exhales, the point of interest stays on the feeling of the breath entering into and leaving the body. Thoughts will continually rise up, but the trick is not to chase them or push them away. Simply observe them and gently return the attention to the breath.

In essence, mindful meditation is a gift, a device that unlocks the door to the existing second. It's a gateway to a life lived honestly, in which every 2nd is skilled deeply, and the richness of now may be absolutely preferred. In the tremendous expanse of time, the existing is all there surely is. By harnessing the strength of aware meditation, one no longer handiest complements reputation and self-attention but additionally learns to stay within the best time that really topics: the prevailing.

Unlocking The Subconscious

To truely grasp the mind and benefit greatness, it turns into vital to understand and harness the strength of the subconscious.

Picture an iceberg. The visible thing above the water represents conscious thoughts and instant recognition. However, the bigger, submerged part represents the subconscious. While it's far hidden from direct view, its sheer length technique it has a large effect at the entire iceberg's direction and balance.

Take, for example, Leo. Outwardly, he became a a fulfillment businessman with the whole lot seemingly going for him. Yet, he often felt unfulfilled and burdened. Despite no longer information why he felt this manner, the solution lay deep inner his subconscious. As a toddler, he had overheard conversations about how wealth did not deliver happiness. This concept, embedded in his unconscious, influenced his feelings approximately his achievements, despite the fact that he wasn't consciously aware about it.

The subconscious thoughts acts as a storehouse for past stories, instructions determined, beliefs obtained, and traumas persisted. It is continuously "on," even if snoozing, and works tirelessly to way facts, find styles, and make experience of the world.

So, how can one delve into the subconscious and use it as a device for increase?

The first step is reflection. By setting apart quiet moments to ponder, you'll start tuning in to underlying emotions, beliefs, and dreams. Journaling may be an powerful tool for this, allowing thoughts and feelings to waft freely onto paper, revealing insights that won't be obvious in the rush of each day life.

Dream assessment gives a few other window into the unconscious. Dreams frequently function a playground for the unconscious to discover unresolved problems, goals, and fears. By recalling and deciphering desires, insights into underlying drives and blockages may be exposed.

Furthermore, hypnotherapy has received recognition as a manner to right now communicate with the subconscious. By inducing a kingdom of deep rest, hypnotherapists can pass the aware mind and get right of access to the unconscious.

This technique can be used to find out deep-seated beliefs or perhaps to implant new, top notch affirmations.

Another approach is visualization. By growing colourful intellectual pics, it is feasible to talk with the subconscious in its very private language. If someone wishes to gain self perception, as an example, they'll be able to visualize themselves reputation tall, speakme without a doubt, and exuding self belief. Over time, these pics can effect the subconscious, which in flip affects behavior within the actual international.

The unconscious is also particularly receptive to affirmations. These are fantastic statements that, while repeated, can reshape ideals and attitudes. For instance, in desire to harboring a belief that "I am no longer right sufficient," which may additionally moreover additionally stay in the unconscious because of beyond research, one could time and again verify, "I am succesful and nicely worth." Over time, this may overwrite the old notion.

Understanding and leveraging the unconscious isn't always a short repair. It's a journey, requiring staying power, willpower, and regularly a whole lot of introspection. However, the rewards are profound. By aligning the subconscious with conscious desires and desires, it becomes a powerful excellent friend. Instead of unknowingly operating in the direction of one's objectives because of hidden fears or beliefs, the unconscious can be harnessed to propel earlier, making sure that actions, emotions, and goals are in harmony.

In the search for greatness, whilst abilities, records, and out of doors elements play a role, actual mastery starts offevolved inner. And within the thoughts, the subconscious holds the crucial aspect. By unlocking its secrets and techniques and tapping into its energy, one can't handiest understand the deeper elements of oneself but also harness them for top notch increase and success.

Chapter 4: The Power Of Positive Affirmations

Words personal electricity. They shape mind, mould perceptions, and feature an impact on actions. But what if there was a manner to intentionally pick out out and repeat words that might bring about preferred changes within and spherical oneself? This transformative technique exists, and it is called the workout of satisfactory affirmations.

At the coronary coronary heart of an affirmation is a easy but profound concept: through consciously deciding on, repeating, and believing in a brilliant announcement, you'll be capable of result in first-class modifications of their emotions, conduct, and truth.

To higher understand this, don't forget Sarah. She grew up paying attention to statements like "Money does not grow on timber" and "We're not the sort who get fortunate in existence." Unknowingly, the ones statements have grow to be a part of her belief system, leading her to be cautious of monetary opportunities and generally pessimistic. However, whilst she learned about tremendous affirmations, she began out repeating to herself: "I am deserving of abundance" and "Luck and prosperity come obviously to me." Over time, she observed a shift. Opportunities regarded to move again her way, and her financial state of affairs progressed.

So, how does a clean repetition of phrases create such powerful alternate?

The mind, wondrous and adaptable, operates based mostly on styles and behavior. When a perception, belief, or emotion is bolstered over and over, it strengthens neural pathways associated with that idea. Positive affirmations, while used efficaciously, can function a device to set up new, favorable neural pathways, essentially rewiring the brain for success, happiness, and positivity.

However, for affirmations to artwork, a few key ideas want to be accompanied:

1. Specificity: Affirmations must be clean and precise. Instead of pronouncing "I need to be happy," one would possibly say, "I discover pleasure in the small moments of each day."

2. Present Tense: To make affirmations effective, they ought to be stated as though they may be already actual. "I am becoming more assured each day" is greater powerful than "I becomes greater confident."

3. Repetition: For any new dependancy or belief to take root, repetition is high. The identical goes for affirmations. They need to be repeated frequently, preferably at specific times of

the day, like upon waking up or in advance than drowsing.

four. Emotion: Simply pronouncing the terms isn't sufficient. Feeling the emotion in the decrease lower back of them, in reality believing them, is what gives them power.

5. Positive Wording: Affirmations ought to be framed genuinely. Instead of "I am not afraid," a more effective confirmation will be "I am brave."

To further recognize the ability of affirmations, test Max, a younger athlete. Despite his expertise, he frequently faltered in vital moments due to nerves. Then he began the use of the affirmation: "I am calm and targeted below stress." He repeated it each day, visualized himself succeeding, and felt the feelings tied to the words. Over time, his normal overall performance in immoderate-stakes moments improved dramatically.

The realm of high-quality affirmations extends beyond private achievements. They may be used to decorate relationships ("I communicate with love and knowledge"), health ("My frame is strong and recovery"), or maybe talents ("I am constantly improving and mastering my craft").

But, as with all device, affirmations aren't magic. They require dedication. Saying an confirmation for an afternoon or and then leaving behind it may not deliver preferred outcomes. However, at the same time as paired with motion and actual notion, they end up an excellent strain, guiding the subconscious mind in the path of a preferred truth.

Positive affirmations provide a bridge between the cutting-edge self and the desired self, between modern-day reality and a dreamt reality. They are seeds. When planted in the fertile floor of the unconscious, nurtured with perception, and watered with repetition, they may become powerful timber, reshaping landscapes and growing new horizons.

Cognitive Behavioral Techniques

It's now common for terrible mind to become constant partners, casting shadows over desires and aspirations. But what if there has been a way to confront the ones terrible mind, assignment them, and replace them with empowering ideals? Enter Cognitive Behavioral Techniques, a difficult and fast of equipment designed to reshape the concept processes.

At its center, Cognitive Behavioral Techniques, often shortened to CBT, is ready recognizing patterns of bad thinking and then systematically difficult and changing them. These techniques are rooted inside the records that mind, emotions, and behaviors are interconnected. A horrible belief can cause awful feelings, which in turn can reason counterproductive behavior.

To illustrate, don't forget the case of Alex. Alex struggled with public speaking. The mere idea of addressing a difficult and rapid could spiral him proper proper into a whirlwind of horrible self-speak: "I'll reduce to rubble," "People will laugh," or "I'm no longer accurate sufficient." These thoughts brought about emotions of tension, and consequently, he ought to each keep away from public speakme engagements or fumble via them, similarly reinforcing his perception in his very personal inadequacy.

The starting point of CBT is attention. Before any change can get up, one have to become aware about the terrible idea styles. In Alex's case, recognizing that he emerge as trapped in a loop of terrible self-speak have end up the first step.

Once those styles are diagnosed, the next step is hard those ideals. This includes thinking the concept of such mind. Is there any evidence to manual the concept? Are there opportunity reasons or views? For Alex, even as he significantly examined his beliefs, he found out that he had efficiently spoken in the the front of groups inside the past, and no longer every person changed into anticipating him to make a mistake. This popularity weakened the maintain of his terrible beliefs.

The final step is reframing. This method changing the identified horrible concept with a greater effective or unbiased one. Instead of "I'll lessen to rubble," Alex began out telling himself, "I'm organized and will do my first-rate." Over time, this new idea pattern started out to replace the vintage, important to improved self assurance and reduced anxiety.

CBT is not quite plenty mind; it also encompasses behaviors. Often, bad belief styles motive avoidance behaviors. By confronting and converting the behaviors, it's miles possible to further solidify the present day, tremendous belief patterns. For Alex, this supposed actively seeking out opportunities to speak in public, on every occasion reinforcing his new notion in his abilties.

Let's take a look at some other instance. Mia continually believed she emerge as lousy at math. Every time she encountered a hard trouble, her instant idea became, "I can't do that." This concept delivered about emotions of frustration, and she or he or he should often give up without attempting. After mastering about CBT,

Mia started out to challenge this perception. She requested herself, "Have I normally been bad at math?" The solution have emerge as no. She had confronted and conquer math challenges within the beyond. She started reframing her belief to, "This is difficult, but I can determine it out with some strive." Gradually, her attitude inside the path of math changed, and she or he started out tackling issues with dedication in location of melancholy.

It's critical to word that CBT isn't approximately denying or suppressing traumatic conditions. It's approximately viewing them in a balanced manner, unfastened from the distortions of bad concept patterns. It's additionally a ability that improves with workout. Over time, the technique of identifying, hard, and reframing becomes more automated, main to a greater amazing and resilient mindset.

Cognitive Behavioral Techniques provide a roadmap out of the maze of terrible wondering. By confronting and reshaping unhelpful beliefs, it turns into viable to chart a route in the direction of growth, achievement, and achievement. The mind, as quickly as an opponent, turns into an amazing friend, helping and propelling toward greatness.

Memory Palaces: Building Intellectual Fortresses

Memory is an astonishing device, pivotal now not best for recalling past evaluations, but furthermore for learning and honing capabilities. But as powerful as it's miles, reminiscence can now and again seem elusive. Everyone has felt that frustration while a selected reality or name hovers really out of reap. However, there may be a method, with roots conducting decrease once more to historic times, that may remodel the way one remembers: the Memory Palace.

The concept of the Memory Palace, moreover referred to as the Method of Loci, hinges on the human brain's splendid ability to hold in mind locations and spatial records. It includes growing a intellectual accumulate of a acquainted region and the use of it to maintain facts in a mounted manner.

Imagine a adolescence domestic. Most people can with out problems navigate their manner thru it in their minds, recalling each room, the furniture, and unique precise data. A Memory Palace taps into this innate spatial reminiscence, turning every room or place proper proper into a 'storage bin' for distinctive facts.

For instance, take into account the project of remembering a list of random gadgets: a pineapple, a bicycle, a ebook, and a telescope. Using the Memory Palace technique, one must accomplice every item with a specific place in the decided on familiar placing. The the front door should have a pineapple wreath, the living room might be occupied thru someone the use of a bicycle, the kitchen table is probably stacked with books, and a telescope might be peering out of a bedroom window. When looking for to maintain in thoughts the listing, walking mentally via the residence and looking each scene should carry once more the gadgets in vivid detail.

Historically, this method emerge as valuable. In historical instances, at the same time as paper became a highly-priced and oral shows had been paramount, scholars and orators used Memory Palaces to maintain in mind huge amounts of statistics. One famous tale is of the Greek poet Simonides, who controlled to consider the names of all of the visitors at a dinner party he attended after the roof collapsed, surely because he remembered in which all people have been seated.

However, the Memory Palace isn't always best a relic of the past. It has sensible packages today. Students can use it to consider dates for facts exams or steps in complicated equations. Professionals can lease the method to endure in mind factors in a presentation or data for an important assembly.

Another modern-day edition may be seen in competitive reminiscence championships. Participants frequently need to don't forget decks of playing playing cards or prolonged strings of numbers. Some of the best competitors employ Memory Palaces, turning each card or quantity right right into a colorful picture and placing it in a intellectual place. This way not exquisite aids take into account but moreover makes the act of memorizing greater engaging and fun.

A actual-life instance showcasing the effectiveness of this approach includes a journalist who determined to delve into the vicinity of aggressive memorization. With no earlier experience, however with the assist of Memory Palaces and first rate mnemonic devices, he went from being a amateur to winning the U.S. Memory Championship in a short span.

The creation of a Memory Palace is a customized undertaking. Each man or woman's palace is molded by using the usage of their reviews, making it specific. It's a dynamic tool, growing as wanted. New rooms or places may be added as one becomes extra cushty and wants to bear in mind greater records.

Moreover, the vividness is top. The brain tends to consider unique, funny, or weird photos better than mundane ones. So, the greater progressive the establishments in the Memory Palace, the more potent the recall.

The Memory Palace is a testomony to the enduring strength of historical know-how. It's a bridge among the past and the triumphing, proving that a few system, irrespective of how antique, remain applicable. By constructing and the use of the ones highbrow fortresses, everyone can launch a brand new diploma of brilliance, turning the every so often elusive streams of memory into concrete, navigable pathways.

Chapter 5: Harnessing Flow States

Everyone has professional those magical moments while time seems to disappear. Hours experience like mins, and the mind and body paintings collectively seamlessly. Everything falls into region, whether or not or no longer or no longer it is writing a bit of tune, gambling a sport, or perhaps solving a complicated hassle. This is a "glide state," generally called "being in the quarter." It's a place where in productivity and creativity skyrocket.

A waft country is a highbrow nation wherein someone is definitely immersed in an activity, major to a heightened experience of reputation and involvement. It's a fusion of movement and focus wherein distractions fade away, self-attention diminishes, and normal performance peaks.

The idea of go with the flow isn't new. It has been described in severa cultures for centuries. However, the term "drift" have become coined within the 1970s with the aid of the use of way of a psychologist named Mihaly Csikszentmihalyi. After extremely good research, he concluded that human beings are happiest when they may be in a country of go with the flow. This u.S., he determined, is characterised via severa elements: clean dreams, instant remarks, a stability amongst potential and mission, and a merging of movement and attention.

To illustrate, don't forget a musician gambling a hard piece. They understand the purpose (to play the piece efficaciously), get right away feedback (the tune they produce), and find out a balance among their capacity and the assignment of the piece. As they play, their movements and attention merge; they lose themselves inside the tune. This is float.

Another instance is an athlete all through a large game. The aim is obvious, the comments is the overall overall performance, and there's a stability amongst their education and the game's challenge. They may additionally later describe the game as a blur, or experience as despite the fact that they had been really "going thru the motions" regardless of gambling notably. Again, that is float.

So, how does one benefit this coveted united states of america? There are high quality situations and practices that might make getting into a drift country much more likely:

1. Challenge-Skill Balance: This is likely the most critical circumstance for waft. The project reachable need to be neither too easy nor too tough. If it's miles too clean, boredom gadgets in; if it's miles too tough, anxiety takes over. The candy spot is in which the task gives simply the right

quantity of task for the character's talent level.

2. Clear Goals: Having smooth goals for the undertaking permits to offer course and reason. Knowing what is expected and what one is trying to acquire can help hold the focal point sharp.

3. Eliminate Distractions: A cluttered or distracting surroundings ought to make it difficult to recognition. It may be beneficial to set up a area specifically for strolling or practicing, free from vain distractions.

4. Immediate Feedback: Knowing how well one is doing in actual-time can help to regulate and refine moves. This remarks loop is vital for retaining the go with the go with the glide u.S..

5. Deep Concentration: Before stepping into a go with the flow nation, it regularly calls for a length of deep focus.

This can also recommend dedicating a selected amount of uninterrupted time to the task.

6. Control: Feeling on top of things over the project can inspire a glide united states. This doesn't mean the entirety is probably predictable, however alternatively that one feels organized to address the traumatic situations that rise up.

By data and cultivating these situations, it becomes extra feasible to harness go together with the glide states, maximizing each productiveness and creativity. It's like unlocking a superpower, wherein the thoughts and frame carry out at peak universal performance. And at the same time as it'd appear elusive, with exercising and purpose, engaging in glide can end up a more commonplace part of one's lifestyles. Whether an artist, an athlete, or someone clearly on the lookout for to make the most out in their workday, tapping into this "quarter" can result in terrific results.

Biohacking Your Brain

In the age of innovation, in which era and biology intersect, there's a developing motion called "biohacking." Biohacking, at its center, is the exercise of converting one's biology and physical surroundings to optimize health, nicely-being, and usual performance. When it involves the brain, biohacking specializes in enhancing cognitive abilties, which incorporates memory, hobby, and creativity.

With the mind being the command center of the body, it is critical to hold it in only circumstance. Just as one would possibly probably song a musical device for the wonderful sound, biohacking allows for the best-tuning of the mind to acquire pinnacle ordinary performance. However, while the concept can also moreover sound futuristic, a whole lot of those strategies are grounded in ancient practices or smooth technological information.

Here are a few safe, era-subsidized techniques to biohack the thoughts:

1. Dietary Changes: Nutrition plays a high-quality role in mind fitness. Foods wealthy in omega-three fatty acids, antioxidants, and positive nutrients and minerals have been confirmed to assist cognitive features. For instance, fatty fish like salmon and walnuts are appropriate assets of omega-3s, which guide mind cellular conversation.

2. Intermittent Fasting: This consists of biking amongst intervals of consuming and fasting. Studies have showed that intermittent fasting can promote brain health with the aid of reducing oxidative

strain, irritation, and promoting the manufacturing of mind-derived neurotrophic element (BDNF), a protein that helps mind feature.

3. Brain Training Games: While the mind isn't a muscle, it could although benefit from a excellent exercise. Certain games and puzzles, like Sudoku or chess, assignment the thoughts and assist beautify numerous cognitive abilties.

four. Mindfulness and Meditation: Regular meditation now not most effective reduces strain but can also increase gray rely within the mind, which is worried in muscle control, sensory perception, and preference making.

five. Aerobic Exercise: Physical hobby, specifically aerobic exercising, has profound benefits for the mind. It can boom the dimensions of the hippocampus, the part of the brain concerned in memory and analyzing. Consider the tale of an aged community that took up synchronized swimming. Over time, many people stated sharper reminiscence and quicker reasoning skills.

6. Sleep Optimization: Sleep isn't only a time for relaxation; it is also while the thoughts approaches information from the day, office work recollections, and preservation itself. Ensuring a normal sleep time table and growing a conducive sleep environment can do wonders for cognitive skills.

7. Limiting Blue Light Exposure: Modern life is full of shows, from smartphones to computer systems. The blue light emitted from the ones gadgets can interfere with the producing of melatonin, a hormone liable for sleep. Reducing display time earlier than bed or using blue mild filters can help make sure

an exceptional night time time's sleep, reaping blessings mind fitness.

eight. Nootropics: These are natural or artificial compounds that would decorate cognitive capabilities. Some not unusual natural nootropics encompass caffeine and L-theanine, often located in tea. However, it's far critical to approach nootropics with warning and normally visit a healthcare professional earlier than beginning any new supplement.

9. Environmental Adjustments: The environment play a big function in how the mind features. A litter-unfastened, properly-lit surroundings with minimal distractions can enhance popularity and productivity.

10. Learning New Skills: Picking up a modern interest or talent, like a musical device or a modern day language, annoying conditions the brain and

continues it active, that could purpose improved neural connections.

In conclusion, biohacking the mind consists of a aggregate of current generation, historic practices, and now and again clearly commonplace enjoy. Whether it's tweaking one's weight loss plan, incorporating meditation, or absolutely ensuring a exceptional night time's sleep, there are numerous strategies to optimize the thoughts's capabilities. As with all things, it's critical to listen to at least one's body, be affected person, and take a balanced method to achieve the top notch outcomes.

Leveraging the Pareto Principle

In the world of general overall performance and effectiveness, the Pareto Principle stands tall as a beacon of facts. Originating from an remark with the aid of the Italian economist Vilfredo Pareto, this principle indicates that type of 80% of outcomes or outcomes come from clearly 20% of motives or movements.

Let's think about a lawn. Imagine a full-size garden full of lots of plant life. Now, assume handiest a small detail, say 20% of those vegetation, produce 80% of the garden's general flora and prevent end result. This lawn serves as a metaphor for masses situations in existence in which a minority of motives are accountable for a majority of consequences.

Consider a agency setting. In many agencies, it is frequently discovered that 20% of the clients account for 80% of the general income. Similarly, in private existence, it might be discovered that 20% of duties or efforts lead to eighty% of the accomplishments or satisfaction.

Understanding and applying the Pareto Principle can also have profound affects on productivity and success. Here's how it could be leveraged:

1. Identify the Critical 20%: Begin with the useful resource of the usage of list duties, goals, or sports. Then determine which of them yield the very excellent rate or consequences. By concentrating efforts on these essential responsibilities, the output may be substantially superior.

2. Eliminate or Delegate the Less Productive 80%: Once the vital 20% is diagnosed, affirm the remaining responsibilities. Some of these may not be essential and may be removed, at the same time as others may be delegated or outsourced.

For example, a eating place owner might discover that 20% of the dishes on the menu account for eighty% of the income. Instead of spreading assets skinny, the proprietor need to attention on perfecting those few dishes and possibly simplify the menu.

three. Prioritize Tasks: With a clean understanding of what obligations or movements are maximum valuable, prioritize them. Make excessive quality they get the most interest and assets.

Imagine a scholar analyzing for an exam. If 20% of the subjects are possibly to cover 80% of the questions, the pupil should make certain those subjects are nicely-understood first.

four. Re-take a look at Regularly: The vital 20% would possibly likely exchange over the years. Therefore, it's far beneficial to re-study duties and goals periodically to make certain alignment with the Pareto Principle.

5. Apply the Principle in Multiple Areas: The beauty of the Pareto Principle is that it could be completed almost

everywhere – from time manage and commercial enterprise agency to non-public relationships and health.

Consider health and health. It might be decided that 20% of exercise sports activities bring about eighty% of the popular health effects. Knowing this, one have to consciousness extra on those particular workout workouts to gain better effects.

6. Avoid the Trap of Perfectionism: Recognizing that a substantial majority of outcomes can come from a minority of efforts can free one from the entice of perfectionism. Instead of seeking to do the whole thing perfectly, it's about doing the proper subjects excellently.

7. Boost Decision Making: When confronted with a plethora of picks, the Pareto Principle can assist in preference-making. By asking which alternatives will

yield the best effects, you may however make extra informed and green alternatives.

The Pareto Principle is more than handiest a precept or idea; it's a guide to streamlined performance and maximized outcomes. By statistics that now not all efforts are created identical and that a minority can motive a majority of effects, humans can harness this principle to reap more with a whole lot a whole lot much less. In a international packed with endless duties and distractions, leveraging the Pareto Principle may be the important component to accomplishing greatness in any company.

Chapter 6: Embracing Constructive Discomfort

Imagine standing at the brink of a enormous and deep pool. At one side, wherein the water is shallow, one feels constant, feet touching the ground and head properly above the water. This is acquainted, comfortable, and predictable. But at the opportunity element, in which the waters run deep, there may be a sense of unease. There's a task, a chance, and an unknown intensity. It's the aspect that guarantees boom but wishes braveness.

This pool symbolizes lifestyles. The shallow aspect represents the consolation sector - a area that feels warmth, acquainted, and steady. The deeper give up? It suggests constructive discomfort, in which increase, reading, and evolution stay.

Understanding the Comfort Zone

It's a herbal human instinct to are searching out comfort. It is in which exercise exercises reside and predictability reigns perfect. But an excessive amount of time in this vicinity can purpose stagnation. The boom curve starts to flatten, and the brilliance that when shone vibrant starts to dim. After all, wood do no longer increase in comfortable floor; they acquire their towering heights through wrestling with the winds and weathering storms.

The Magic of Constructive Discomfort

Stepping out of the comfort vicinity and embracing best pain can result in profound personal and expert growth. It is in this space that creativity sparks, resilience strengthens, and competencies sharpen.

Consider a younger infant studying to revel in a bicycle. The initial wobbles, the fear of falling, and the hesitations are all part of the discomfort. But with each push of the pedal and each small distance included with out manual, the child grows more potent, more assured, and in the end masters the artwork of cycling.

Similarly, whilst an person comes to a decision to analyze a modern-day language, the first steps can be daunting. The surprising sounds, the complex grammar regulations, and the task of protective a verbal exchange can all appear overwhelming. But with time, exercising, and perseverance, the as speedy as-remote places language turns into a fluent shape of expression.

Navigating the Landscape of Discomfort

1. Start Small: One does not need to dive deep into the pool proper away. Start via dipping a toe, then a foot, and frequently wade into deeper waters. Whether it's far selecting up a modern

interest, trying a superb form of food, or speakme in public for the number one time, every step counts.

2. Shift Perspective: Instead of viewing ache as a daunting assignment, see it as an opportunity. An possibility to enlarge, to study, and to reinforce.

three. Seek Support: Venturing into the unknown can be heaps tons much less intimidating with a manual tool. Whether it is buddies, family, or colleagues, having a person to proportion memories, searching for recommendation, or honestly provide terms of encouragement should make all of the distinction.

four. Celebrate Small Wins: Every milestone, no matter how minor it may appear, is a testament to development. Celebrating these wins fuels motivation and propels one further into the journey of positive pain.

five. Remember the 'Why': Whenever doubts creep in or worrying conditions seem too overwhelming, remind oneself of the 'why'. Why did one pick out out to step out of the consolation area? Why is this adventure essential? The 'why' is a powerful anchor that could ground and guide through the roughest of seas.

6. Stay Adaptive: The landscape of pain is ever-changing. What seems hard these days could probable become 2d nature day after today. The secret is to live adaptive, flexible, and open to alternate.

Constructive Discomfort as a Catalyst

Embracing quality discomfort isn't approximately looking for soreness for its very very own sake. It's about recognizing that ache, at the same time as approached with the right thoughts-set and system, can be a effective catalyst for increase. It can push limitations, shatter limits, and propel to awesome greatness.

In the narrative of lifestyles, consolation is probably the creation, but constructive pain is in which the plot thickens, characters expand, and the tale takes memorable turns. Embrace it, examine from it, and use it as a stepping stone to gain the pinnacles of private and expert achievement.

Taming the Technology Beast

From smartphones to laptops and capsules, technological wonders have modified the way paintings is finished, relationships are built, and enjoyment is ate up. But with all its blessings, if left unchecked, this beast can outcomes reason distraction, reduce productiveness, and scouse borrow treasured moments of lifestyles.

The Double-Edged Sword of Technology

Technology is sort of a coin with elements. On one side, it gives a global of facts at fingertips, simplifies complex obligations, and keeps anybody related. However, the opportunity difficulty can bring about infinite scrolling, hours wasted on beside the point content material material, and a non-stop usa of distraction.

For instance, don't forget the cell smartphone. It's a tool which can assist navigate to a modern-day region, video name loved ones miles away, or even order meals on the touch of a button. Yet, the identical tool can bring about hours out of place in social media, a consistent barrage of notifications, and a reluctance to disconnect.

Strategies for Manipulation, Not Distraction

The purpose isn't always to take away generation however to tame it, to make it serve specific features as opposed to allowing it to be the draw close.

1. Set Clear Boundaries: Determine specific instances for checking emails, social media, and extraordinary apps. By restricting those sports to set intervals, it will become less tough to attention on responsibilities with out steady interruption. For instance, a person also can decide to simplest take a look at their social media money owed for 20 mins after lunch in choice to sporadically in the course of the day.

2. Use Technology to Block Technology: There are apps designed to help lessen display time or block distracting net websites. By putting in place the ones apps, the temptation to wander off into the virtual realm can be curtailed. For instance, an app might

probably block get right of entry to to certain web sites in a few unspecified time inside the future of paintings hours, ensuring interest is maintained.

3. Notifications – The Silent Intruder: Each time a device pings, interest is diverted. It's useful to reveal off non-essential notifications. By simplest permitting important signs, it reduces the normal pull to observe the show.

4. Physical Boundaries: Sometimes, the best answers are the handiest. If running on a task that calls for deep interest, putting the smartphone in every other room can work wonders. Out of sight frequently way out of thoughts.

5. Designated Tech-Free Times: Just as there are set times for using era, there need to moreover be moments even as monitors are intentionally placed away. This can be in the course of food, the first

hour after waking up, or the final hour in advance than bed.

6. Mindful Consumption: Before putting in an app or a internet internet site, take a 2d to ask, "Is this necessary? What's the reason?" By being intentional with technology use, it prevents aimless wandering in the virtual worldwide.

7. Quality over Quantity: Instead of in search of to be present on each social media platform or downloading each trending app, choose a select out few that virtually upload price. It's better to deeply have interaction with some structures than to spread oneself too thin at some stage in many.

From Beast to Ally

By enforcing those strategies, the reference to era shifts. It transforms from a beast that constantly desires interest to a valuable pleasant buddy that complements lifestyles.

Imagine a student making ready for a big test. Instead of being distracted thru every notification, they've got set their tool to "Do Not Disturb." The handiest app they've got open is one that plays calming instrumental music, supporting them reputation. This pupil has successfully tamed their era beast, making it a powerful tool for his or her achievement.

The technology beast isn't going away. Its presence is ever-developing, with new gadgets and systems rising frequently. But with the aid of setting easy barriers, being intentional, and the use of generation mindfully, it can be tamed. The reason is to ride this beast, guiding it in useful instructions, in desire to being trampled via its relentless march.

Chapter 7: The Gratitude Advantage

In a international that frequently highlights what's missing, there exists a clean but transformative exercising: gratitude. At its center, gratitude is the acknowledgment and appreciation of the fantastic elements of life. Whether for the air one breathes or for a kindness received from a chum, gratitude gives a lens thru which the arena can be visible in a brand new slight.

Understanding Gratitude

Gratitude is not pretty lots pronouncing "thanks." It is a deeper recognition of the coolest, each big and small, that permeates everyday existence. By focusing on remarkable factors, a shift occurs in how one perceives and reacts to the arena.

For instance, take into account humans caught in web site traffic. One individual specializes within the out of place time, the inconvenience, and the frustration. They experience stressed and stressful. The distinctive character sees this pause as an opportunity. They are thankful for the greater time to pay attention to tune, reflect, or simply breathe. The scenario is the equal, but the perspectives are really high-quality. The latter man or woman has tapped into the energy of gratitude.

Gratitude and the Brain

Recent research have demonstrated that gratitude can literally change the mind. When a person regularly practices gratitude, there are increased ranges of hobby inside the hypothalamus, a essential part of the brain answerable for numerous essential obligations at the side of strain stages and sleep. This multiplied interest can bring about higher sleep, reduced stress, or maybe a more potent immune system.

Furthermore, specializing in fine sports and being thankful can growth the manufacturing of dopamine, a neurotransmitter that rewards the mind with emotions of delight. It's like giving the thoughts a "sense-proper" address on every occasion gratitude is practiced.

The Domino Effect of Gratitude

When gratitude is practiced frequently, it begins offevolved a series reaction of positivity. Here's the way it really works:

1. Improved Mental Health: By recognizing and appreciating the coolest, bad emotions and mind are not noted, main to decreased emotions of envy, resentment, and frustration.

2. Better Physical Health: Those who workout gratitude report feeling healthier, experiencing less ache, or perhaps living longer. This is probable due to the

decreased pressure and prolonged positivity that gratitude brings.

three. Enhanced Empathy: Instead of feeling resentment closer to others, thankful people have a propensity to be greater statistics and compassionate.

4. Boost in Self-esteem: Recognizing one's very own benefits can purpose a more appreciation of oneself and one's achievements.

Practicing Gratitude

Cultivating gratitude is not a complex procedure. It can be as clean as taking a few moments every day to mirror on the positives. Here are some methods to combine gratitude into day by day lifestyles:

1. Gratitude Journal: Every night time, jot down 3 matters to be glad approximately from the day. Over time,

this workout can create a first rate comments loop, schooling the mind to are seeking for out and cognizance on first-class additives.

2. Gratitude Jar: Place a jar in a commonplace vicinity. Whenever feeling grateful for some trouble, write it down and vicinity the notice within the jar. Over time, the jar fills with countless moments of gratitude.

three. Verbal Expression: Take the time to particular gratitude to others. Whether it's far for a small preference or just for being there, letting someone recognize they're favored can be powerful for every the giver and the receiver.

The Story of Anna

To illustrate the transformative strength of gratitude, keep in mind Anna. She confronted worrying conditions at paintings, health concerns, and the general stresses of each day existence. Initially, those worrying situations consumed her, essential to anxiety and disappointment. However, after being added to the idea of a gratitude magazine, she determined to offer it a attempt.

Each night time time, Anna wrote down three subjects she have grow to be grateful for. Some days, it come to be as smooth as "a warmth bed" or "a name from a pal." Over time, Anna started to be aware a shift. She felt lighter, more satisfactory, or perhaps commenced to face demanding situations with a newfound resilience. Anna's mindset on life changed, no longer because of the truth her circumstances did, however due to the fact her consciousness did. She had placed the gratitude advantage.

In give up, on the identical time as stressful conditions and hardships are part of life, the lens thru which they may be taken into consideration ought to make all the distinction. Gratitude gives a powerful lens, one that specializes in the effective, promotes intellectual and physical well-being, and may transform the normal into the excellent. Embracing gratitude isn't just about seeing the glass as half of entire; it's far approximately being thankful that there is a pitcher the least bit.

Nurturing a Growth Mindset

A mind-set, in its most easy form, is a set of beliefs or attitudes about essential elements of the world. People supply numerous mindsets approximately love, success, challenges, and, most importantly, approximately themselves. The growth thoughts-set facilities on the perception that talents and intelligence may be developed with attempt, workout, and staying electricity. In stark evaluation, a hard and rapid thoughts-set assumes that the ones inclinations are innate and unchangeable.

Understanding the Fixed vs. Growth Mindset

Imagine university college students making equipped for an examination. The first student, with a tough and rapid mindset, thinks, "I'm not right at this assignment. I've in no way been, and I in no manner can be." The 2nd pupil, with a growth mindset, thinks, "This is hard, however with sufficient have a examine and workout, I can grasp it."

The distinction in their technique is apparent. The student with a hard and fast mindset feels defeated in advance than even beginning, believing competencies are set in stone. In evaluation, the scholar with a boom thoughts-set sees capability and is conscious that strive plays a essential position in gaining knowledge of any ability or situation.

The Power of 'Yet'

One of the most massive insights from the studies on increase thoughts-set is the power of the word "yet." When faced with a undertaking or setback, including "however" transforms the attitude. "I can't try this" turns into "I can't do this but." This simple word opens up a worldwide of opportunities and underscores the understanding that with effort and time, boom can rise up.

Benefits of a Growth Mindset

Adopting a boom thoughts-set has a ripple impact on severa factors of existence:

1. Embrace Challenges: Instead of avoiding traumatic conditions, they grow to be exciting opportunities for boom.

Each challenge is a stepping stone to a better stage of information or capacity.

2. Persist inside the Face of Setbacks: Failures aren't seen as defining moments however as comments. They come to be gaining knowledge of possibilities, guiding what wishes to be completed in another way subsequent time.

three. Effort as a Path to Mastery: The system turns into as critical, if not extra so, than the prevent end end result. The journey, with all its americaand downs, is wherein authentic mastering and boom appear.

4. Learn from Criticism: Instead of becoming defensive or disheartened via way of comments, it's miles seemed as valuable information that could assist enhance and develop.

5. Find Lessons and Inspiration within the Success of Others: Instead of feeling

threatened thru others' fulfillment, it is used as motivation. The accomplishments of friends become a source of proposal and mastering.

Cultivating a Growth Mindset

Changing one's mindset isn't an in a single day undertaking. It's a adventure. However, with conscious attempt, it's miles feasible to shift from a hard and fast to a boom mindset.

1. Awareness: Recognizing and acknowledging cutting-edge ideals is the first step. By being privy to regular thoughts-set mind when they rise up, you can still mission and reframe them.

2. Value the Process: Focus on the reading method, not surely the final effects. Celebrate small victories and development, notwithstanding the fact that the cease motive isn't always reached right now.

3. Reframe Challenges: See worrying situations as possibilities. Each impediment is a hazard to enlarge and studies some issue new.

four. Use Failure as a Teacher: Instead of being disheartened via setbacks, have a look at them. What can be discovered? How can this guide future efforts?

5. Seek Feedback: Regularly are searching out comments and use it constructively. It's a beneficial device for growth.

Story of Marcus

To in reality understand the transformative electricity of the boom mindset, consider the story of Marcus. Growing up, Marcus struggled with analyzing. He regularly idea, "I'm simply now not a reader. It's no longer in my genes." This notion held him decrease back for years. However, in excessive faculty, a instructor added him to the idea of the growth thoughts-set. With this new attitude, Marcus started out to don't forget that maybe he must decorate.

He started out to put in greater hours, sought assist, and slowly, with numerous attempt, his studying skills advanced. By the time he graduated, not great had he stuck up along collectively together with his buddies, however he additionally advanced a deep love for literature.

Marcus's transformation wasn't because of a few hidden capabilities growing. It have grow to be the end end result of his shift in mind-set. He believed he ought to enlarge, and that belief made all the difference.

Nurturing a boom thoughts-set is greater than just a mental exercise; it's far a profound shift in how one approaches worrying conditions, setbacks, and life in favored. By knowledge that abilities and intelligence are malleable, that they will be honed and advanced, one opens up a global of opportunities. The adventure of increase is probably fraught with challenges, however with the proper mind-set, those challenges grow to be the very stepping stones to greatness.

Chapter 8: Organizing Thoughts For Clarity

Our thoughts frequently race spherical in chaotic styles. Ideas sprout, reminiscences resurface, and imaginations run wild. This unorganized idea technique may be lovely in its spontaneity, however for clean thinking and powerful planning, a greater established method is useful. This is wherein the technique of mind mapping enters.

At its middle, a mind map is a seen illustration of information, radiating outwards from a huge idea. Think of it because of the fact the roots and branches of a tree, in which the principle concept is the trunk, and the associated mind are the branches. This approach gives an intuitive manner to set up and form statistics, making it less hard to understand, preserve in mind, and act upon.

The Basics of Mind Mapping

A mind map normally starts with a considerable concept or scenario rely number positioned within the middle of a web web web page. From this number one concept, important issues or subtopics radiate outwards. These problem topics then department out in addition into smaller, greater particular thoughts or information.

The energy of mind mapping lies in its simplicity. By supplying facts in a seen and interconnected manner, the mind can machine, understand, and remember it greater efficaciously.

Benefits of Mind Mapping

1. Clarity: By laying out thoughts visually, you may see connections and relationships more really. This clarity may be instrumental in hassle-solving, making plans, and brainstorming lessons.

2. Memory Boost: The human thoughts is absolutely willing toward visuals. By changing linear notes into colourful and interconnected diagrams, it is much less complicated to do not forget statistics.

3. Enhanced Creativity: The non-linear nature of thoughts maps encourages unfastened association, essential to more revolutionary thinking and idea technology.

4. Time-saving: Organizing data in a streamlined manner allows for quicker assessment and information.

five. Flexibility: Mind maps can be resultseasily adjusted, making them superb for evolving initiatives and thoughts.

Creating an Effective Mind Map

Crafting a a success mind map includes more than truely drawing traces and writing terms. Here are a few guiding ideas:

1. Start in the Center: Begin with the precept concept or problem rely range within the middle. This offers the mind freedom to unfold out in all commands.

2. Use Colors: Different shades can represent different topics or display relationships. They moreover make the map more engaging and much less complicated to endure in thoughts.

three. Embrace Images: A image is simply worth 1000 terms. Wherever feasible, use symbols, icons, or drawings to symbolize mind.

four. Limit Word Count: For every branch or idea, use single phrases or quick terms. This maintains the map easy and clean to study.

five. Connect Ideas: Show relationships amongst unique branches or nodes. This interconnectedness is what offers a thoughts map its electricity.

6. Keep It Flowing: Allow thoughts to conform glaringly. If a cutting-edge-day predominant concept emerges, don't hesitate to branch off in a brand new direction.

Story of Elena

To illustrate the realistic utility and advantages of thoughts mapping, bear in mind the tale of Elena. Elena had a ardour for baking and dreamed of starting her personal bakery. But every time she sat down to devise, she felt beaten via the vastness of the undertaking. From price range to recipes, from location scouting to advertising – there have been a plethora of factors to undergo in mind.

One day, Elena became introduced to the idea of thoughts mapping. She determined to offer it a strive. In the middle of a large piece of paper, she wrote "My Bakery." From there, she started out to branch out. One big branch changed into titled "Finances," which similarly branched out into "Initial Investment," "Monthly Expenses," "Pricing," and so on. Another important branch come to be "Menu," which branched into "Cakes," "Pastries," "Breads," and extra.

As Elena's thoughts map grew, so did her clarity. What as quickly as regarded like an insurmountable challenge now appeared greater possible. She want to peer connections, prioritize responsibilities, and formulate a clearer direction of action.

Within a yr, together together with her mind map as her guide, Elena efficaciously opened her bakery. The seen readability supplied by way of the use of the mind map had transformed her dream into fact.

Mind mapping is a useful tool for organizing thoughts, boosting memory, and improving creativity. In a worldwide overflowing with records, the ability to sift thru the noise, find out connections, and reputation on what honestly subjects is crucial. Whether for personal projects, instructional studies, or professional plans, thoughts mapping offers a pathway to clarity and achievement.

Chapter 9: Social Connection The Neurological Boost

From the sunrise of human statistics, humans have thrived within the organisation of others. There's a very unique magic within the shared laughter with a friend, the comforting consist of of a loved one, or the smooth act of speakme with every exceptional person. It's greater than just emotional properly-being; strong social ties without delay have an effect on intellectual health. This bankruptcy delves into the technological expertise and testimonies in the lower lower back of the neurological benefits of social connection.

Human beings, by the usage of nature, are social creatures. Across awesome cultures and eras, the charge of network, family, and friendship has been recognized and celebrated. However, it is no longer quite lots way of life or sentimentality. There's a wealth of evidence pointing to the tangible neurological benefits of having strong social ties.

The Neurological Impacts of Social Connections

1. Release of Positive Hormones: When humans connect with others, the thoughts releases oxytocin, every so often known as the "love hormone" or "bonding hormone." This hormone plays a crucial role in building agree with and forming emotional bonds. Additionally, engaging in powerful social interactions can bring about the discharge of endorphins, the body's herbal painkillers, which promote feelings of pride and decrease pressure.

2. Reduction in Stress Hormones: Interacting with loved ones can lessen the levels of cortisol, a hormone related to stress. Lower pressure levels aren't simply precise for intellectual nicely-being however moreover have severa fitness

blessings, together with stepped forward immune function.

three. Cognitive Sharpness: Regular social interactions can maintain the thoughts sharp and may even put off cognitive decline as human beings age. Engaging in conversations, debates, and different kinds of social interactions calls for the mind to system records, bear in mind special viewpoints, and formulate responses, all of which make a contribution to cognitive fitness.

4. Elevated Mood: Loneliness and isolation are regularly associated with despair. On the turn issue, common social interactions and robust social assist can growth temper and provide a buffer toward emotions of sadness or depression.

Stories from the Real World

Consider the tale of Liam, a more youthful guy who moved to a contemporary-day town for artwork. The pleasure of the trendy system speedy diminished as he realized that he became a ways from his circle of relatives, friends, and familiar surroundings. The days appeared longer, and the nights had been quiet and lonely. Despite his initial reluctance, he decided to sign up for a neighborhood membership in which humans met to talk approximately books and proportion tales. Slowly, Liam commenced to shape new connections and friendships. With time, not handiest did his mood enhance, but he furthermore felt sharper, extra engaged, and truely happier. He determined out that those tremendous changes had been in big component because of the strong social ties he had began to forge in his new town.

In a few exclusive example, Maria, an aged lady, lived on my own after her youngsters moved to terrific cities. She started out to sense the weight of loneliness, regularly locating herself out of place in mind and recollections. Her neighbor, noticing this, delivered her to a community middle in which senior citizens met, shared recollections, completed video games, and spent time together. The transformation in Maria became apparent. She regarded extra colourful, laughed more regularly, and her reminiscence moreover seemed sharper.

Building and Maintaining Social Connections

Given the easy advantages of social connections, it's crucial to actively nurture them. Here are a few techniques to forge and maintain social ties:

1. Join Clubs or Groups: Whether it's a interest club, a sports activities sports institution, or a network provider corporation, becoming a member of such collectives can provide regular social interactions and open the door to new friendships.

2. Volunteer: Volunteering not most effective offers an opportunity to provide lower back to the community but additionally gives a platform to meet and engage with like-minded individuals.

3. Stay in Touch: In the age of generation, distance is not a barrier. Regular calls, messages, or video chats with loved ones can keep the bond sturdy.

4. Attend Social Gatherings: Even if it might every now and then enjoy like a chore, attending activities, gatherings, or community sports can be an exquisite manner to socialise.

five. Seek Professional Help if Needed: If emotions of loneliness or isolation turn out to be overwhelming, it might be useful to are looking for treatment or counseling.

While contemporary-day lifestyles often emphasizes individual achievements and independence, it's far important to don't forget the age-vintage attention: humans thrive inside the enterprise organization of others. The neurological advantages of social connections are manifold, from the release of superb hormones to progressed cognitive function. It's not just about feeling splendid; it is about being mentally greater healthy and extra resilient. The stories of human beings like Liam and Maria underscore the transformative power of social ties. So, in the adventure of existence, while non-public desires and goals are vital, it is equally essential to walk alongside others, sharing moments, memories, and mutual guide.

Chapter 10: From Manipulation To Mastery

The electricity of the mind is huge. When harnessed, it may shape destinies and redecorate ordinary moments into incredible achievements. However, there may be a pleasing line among controlling and being controlled, among manipulating oneself for proper or for sick intentions. This bankruptcy delves deep into the concept of ethical self-manipulation, guiding human beings on a path toward mastery and personal greatness.

Manipulation, in maximum contexts, includes a terrible connotation. It brings to mind deceit, manage, and ulterior reasons. However, if completed ethically and with pure intentions, manipulating oneself can be a pathway to self-improvement and intention recognition.

Understanding Ethical Self-Manipulation

At its middle, moral self-manipulation is set developing behavior, bodily sports, and mindsets that help in reaching personal and expert desires. It's approximately know-how oneself, recognizing weaknesses, and then tweaking or manipulating perception strategies and actions to transform those weaknesses into strengths.

For example, bear in mind the story of Alex, a professional pianist. He had the abilties and the passion however struggled with level fright. Every time he notion approximately appearing inside the the front of an target audience, his palms may need to sweat, his coronary heart may additionally want to race, and he'd bear in mind all of the topics that would flow wrong. But Alex did not deliver in. He decided to control his fear. Instead of keeping off performances, he started with small shows, the use of visualization techniques to calm his nerves and repeated affirmations to decorate his self perception. With time, Alex transformed his worry right right into a using stress, making him one of the most sought-after pianists in his city.

The Steps to Mastery

1. Self-recognition: Before manipulating any element of oneself, knowledge and accepting cutting-edge-day strengths and weaknesses is important. This step is set honest self-mirrored photograph, searching inward, and evaluating one's right self.

2. Setting Clear Goals: Knowing the popular very last results lets in in charting out a clear direction. Whether it's far reading a capabilities, overcoming a worry, or developing a present day addiction, having easy and plausible goals is essential.

three. Developing Strategies: Once the desires are smooth, it's time to craft techniques or steps that assist in sporting out them. This may contain deciding on up new conduct, letting cross of positive ideals, or attempting to find out of doors assist.

four. Consistent Practice: Mastery isn't always finished in a single day. It's the result of constant efforts, workout, and studying from failures. It's approximately manipulating oneself to live on the right song, even though the path receives difficult.

5. Feedback and Adjust: Regularly comparing one's improvement and being open to feedback can provide treasured insights. Based on the ones, making crucial changes guarantees that the journey remains aligned with the dreams.

www.ingramcontent.com/pod-product-compliance
Lightning Source LLC
Chambersburg PA
CBHW070954080526
44587CB00015B/2309